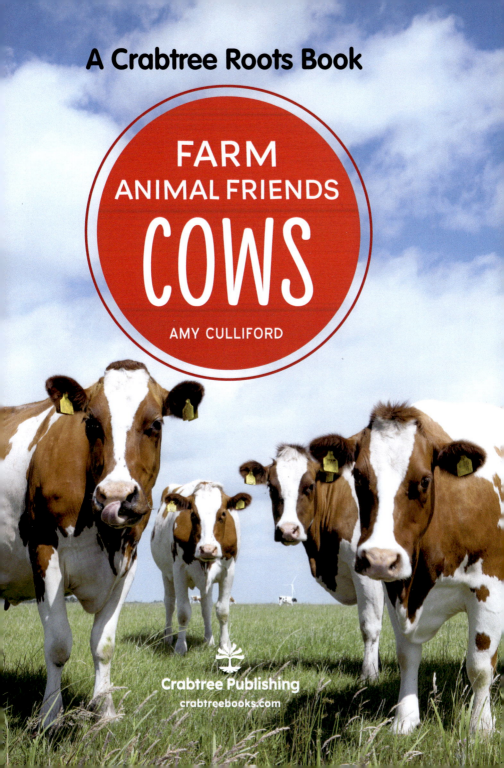

A Crabtree Roots Book

FARM
ANIMAL FRIENDS
COWS

AMY CULLIFORD

Crabtree Publishing
crabtreebooks.com

School-to-Home Support for Caregivers and Teachers

This book helps children grow by letting them practice reading. Here are a few guiding questions to help the reader with building his or her comprehension skills. Possible answers appear here in red.

Before Reading:

• What do I think this book is about?
- *This book is about cows.*
- *This book is about cows that live on farms.*

• What do I want to learn about this topic?
- *I want to learn about all kinds of cows.*
- *I want to learn what colors a cow can be.*

During Reading:

• I wonder why...
- *I wonder why cows have spots.*
- *I wonder why cows say moo.*

• What have I learned so far?
- *I have learned that cows can be different colors.*
- *I have learned that some cows can make milk.*

After Reading:

•What details did I learn about this topic?
- *I have learned that cows can be brown.*
- *I have learned that not all cows make milk.*

• Read the book again and look for the vocabulary words.
- *I see the word **farm** on page 5 and the word **milk** on page 10. The other vocabulary words are found on page 14.*

This is a **cow**.

These cows are on a **farm**.

Some cows are brown

Some cows are black and white.

Some cows make **milk**

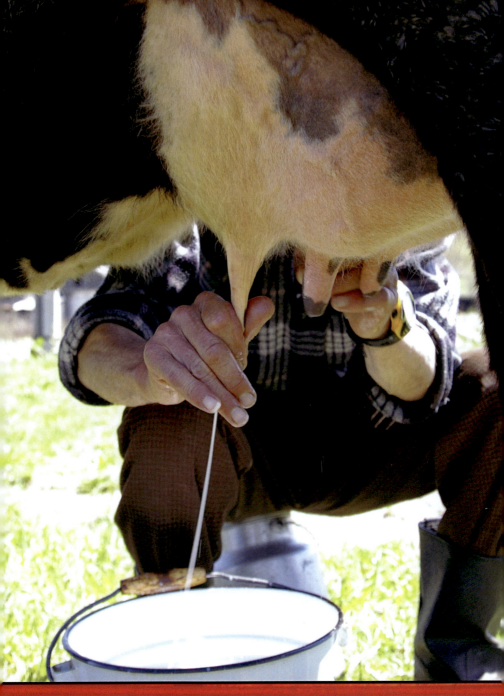

All cows can moo!

Word List

Sight Words

a	black	some
all	brown	these
and	is	this
are	make	white
can	on	

Words to Know

cow

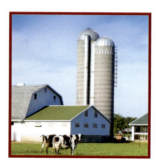
farm

milk

28 Words

This is a **cow**.

These cows are on a **farm**.

Some cows are brown.

Some cows are black and white.

Some cows make **milk**.

All cows can moo!

FARM ANIMAL FRIENDS
COWS

Crabtree Publishing

crabtreebooks.com   800-387-7650

Written by: Amy Culliford
Designed by: Rhea Wallace
Series Development: James Earley
Proofreader: Kathy Middleton
Educational Consultant: Christina Lemke M.Ed.
Photographs: Shutterstock: Clara Bastian:
cover (tl); Sebastian Knight: cover (tr);
Oligo22: cover (b); Anton Havelaar: p. 1; Clara
Bastian: p. 3, 12, 14; Maxy Me: p. 4, 14; L.M.
Dunn: p. 7; SGr: p. 8-9; RedTC: p. 11, 14

Library and Archives Canada Cataloguing in Publication
Title: Cows / Amy Culliford.
Names: Culliford, Amy, 1992- author.
Description: Series statement: Farm animal friends |
    "A Crabtree roots book".
Identifiers: Canadiana (print) 20200382586 |
    Canadiana (ebook) 20200382594 |
    ISBN 9781427134516 (hardcover) |
    ISBN 9781427132468 (softcover) |
    ISBN 9781427132529 (HTML)
Subjects: LCSH: Cows—Juvenile literature.
Classification: LCC SF197.5 .C85 2021 | DDC j636.2—dc23

Published in Canada
Crabtree Publishing
616 Welland Avenue
St. Catharines, Ontario
L2M 5V6

Published in the United States
Crabtree Publishing
347 Fifth Avenue
Suite 1402-145
New York, NY 10016

Hardcover   978-1-4271-3451-6
Paperback   978-1-4271-3246-8
Ebook (pdf)   978-1-4271-3252-9
Epub   978-1-4271-4596-3
Read-along   978-1-4271-3522-3
Audio book   978-1-4271-4595-6

Printed in the U.S.A./112023/PP20230920

Library of Congress Cataloging-in-Publication Data
Names: Culliford, Amy, 1992- author.
Title: Cows / Amy Culliford.
Description: New York : Crabtree Publishing Company, 2021. |
    Series: Farm animal friends : a Crabtree roots book | Includes
    index. | Audience: Ages 4-6 | Audience: Grades K-1 | Summary:
    "Early readers are introduced to cows and life on a farm.
    Simple sentences accompany engaging pictures"-- Provided
    by publisher.
Identifiers: LCCN 2020049774 (print) |
    LCCN 2020049775 (ebook) |
    ISBN 9781427134516 (hardcover) |
    ISBN 9781427132468 (paperback) |
    ISBN 9781427132529 (ebook)
Subjects: LCSH: Cows--Juvenile literature. | Livestock--Juvenile
    literature.
Classification: LCC SF197.5 .C85 2021 (print) | LCC SF197.5 (ebook)
    DDC 636.2--dc23
LC record available at https://lccn.loc.gov/2020049774
LC ebook record available at https://lccn.loc.gov/2020049775